HONEYBEES VISION

RECENT DISCOVERIES

Adrian Horridge

Honeybees Vision Recent Discoveries
ISBN: 978-1-914934-15-5

Text and Graphics: Adrian Horridge
Adrian-Horridge.org

Published by Northern Bee Books 2021
Northern Bee Books, Scout Bottom Farm,
Mythomroyd, Hebden Bridge, HX7 5JS (UK)
www.northernbeebooks.co.uk
Tel: +44 (0) 1422 882751

Designed by: Lynnette Busby
www.whatever.graphics

HONEYBEES VISION

RECENT DISCOVERIES

Adrian Horridge

Contents

Not What You Assume

To most of us, it is obvious that foraging bees see shapes and colours of flowers that they pillage. However, in the nineteenth century a few interested scientists showed otherwise with experiments that are easily repeated. Felix Plateau, son of the mathematician, trained bees to go to coloured paper flowers for a reward of odourless sugar, and found that they returned correctly although he changed the colours and shapes. From similar tests with coloured papers, Auguste Forel, a professor at Zürich, concluded the same.

Carl von Hess and Karl von Frisch

In an excellent book on comparative vision in different animals, Carl von Hess, professor of ophthalmology at Münich, doubted that honeybees have colour vision. His bitter opponent, Karl von Frisch, in a huge effort, trained bees to go to various colours and tested them with the training colour against a square array of papers in 15 different shades of grey. He said that his bees distinguished all colours from all shades of grey, as humans can do, but a careful reading of his long paper reveals that his trained bees could distinguish blue and yellow, but not mid-grey or various green colours from his panel of different grey levels. I found recently that bees do not distinguish grey levels, except by blue content, and they sum together a panel of different colours or greys to give an average value, so von Frisch was fooled by his choice of training and test patterns. However, he became a powerful professor at Münich, with his own journal, and suppressed any objection to his opinion.

Bees have no receptor for yellow, and yellow has little emission
of blue, so this flower was detected by green contrast at its edges.

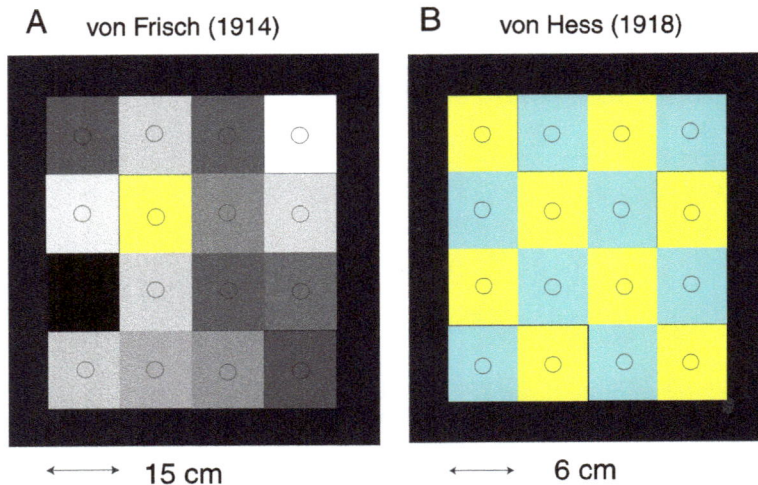

Figure1. A. With the target laid flat, von Frisch trained on a colour, with no alternative, then tested with the colour among 15 different levels of grey. B. When von Hess trained on a blue square of a checkerboard, the bees learned to go to blue, but when trained on a yellow square, they learned nothing about colour, and confused yellow with black. Only one of the little round dishes had the sugar solution, the rest contained water.

Finally, in 1918, Hess trained bees on a 4 × 4 chequerboard of blue and yellow squares (Figure 1B). When trained on blue, the trained bees went to blue in tests, but when trained on yellow, they failed to distinguish yellow from other colours or black. This experiment was repeated in 1973 by Irmgard Kriston, a student of Menzel in Frankfurt. Bees are normally attracted to blue, but when trained on grey or black, her bees confused yellow with black.

Anomalies Unanswered

This early work went unnoticed by English speaking researchers and text book writers, who stayed with the traditional popular opinion of Aristotle that bees distinguish the hues of colours. No-one consolidated the literature, and no-one could produce a new theory. In the 1990's, I was searching for cues that bees detect in black patterns on white paper and came across a curious finding in German by Mathilde Hertz, who had fled to Cambridge and published again in English in 1939. She found that ultraviolet emission from the background white inhibited the detection of black patterns on white paper, a fact that showed trichromatic colour vision in bees to be impossible. There was also a fantastic anomaly discovered in 1979 by Bernhard Ronacher, who trained bees to distinguish between a large grey spot and a smaller black spot, each on a white background with little UV emission. Ronacher found that for each size of the smaller disc, the grey level of the larger disc could be adjusted to make them indistinguishable to the trained bees, although new bees could easily distinguish them.

As shown on the next page when I analysed this with equiluminant colours (Figure 2C) to separate the cues in blue and then in green channels, I found that the bees measured the total blue level in each. They were comparing the blue content of the grey spot against the blue in the greater area of white around the small spot. At the neutral point when they were equal, the change in width of edges of the small target just counterbalanced the total change in blue emission of the grey of the larger spot (See Figure 6, below). A great deal of patience and a huge number of attempts with hundreds of trained bees was essential to discover these unexpected results.

Black patterns on a white background are obviously seen by humans and detected by bees, but black is not a stimulus; there are no black photons. For every object or shape, humans first detect edges and then fill in colours with cues from their three photoreceptor types with peak in red, green and blue. Bees also detect edges that they scan with their green channel, but cannot detect colours of areas. They detect, measure and learn, only the relative intensity of the blue part of the emission. It is not surprising that blue was the first colour to be incorporated into vision, because shorter wavelengths have higher energy and are more likely to find resonating molecules. To discover these simple facts, it was essential to train the blue and the green channels separately in different groups of bees, by the use of training patterns equiluminant to green and then others equiluminant to blue. Further results on colour are later in this account.

A New Start, Measure the Inputs

The visual signals that foraging honeybees detect were discovered by classical methods of scientific analysis. Relative sensitivity of the receptor cells to different colours and intensities were measured with micro-electrodes placed in the compound eye. This defined the receptor contribution. Marked bees that had been trained to distinguish between two large targets with calibrated emission, were tested with suitable targets to discover what essential cues they had learned. They were also tested with versions of the training patterns from which possible cues were omitted. This laborious process defined the feature detectors. Ultraviolet receptors apparently are not connected to memory, and they inhibit perception of blue. With a great variety of targets, this process took years to obtain significant numbers of results to identify all cues.

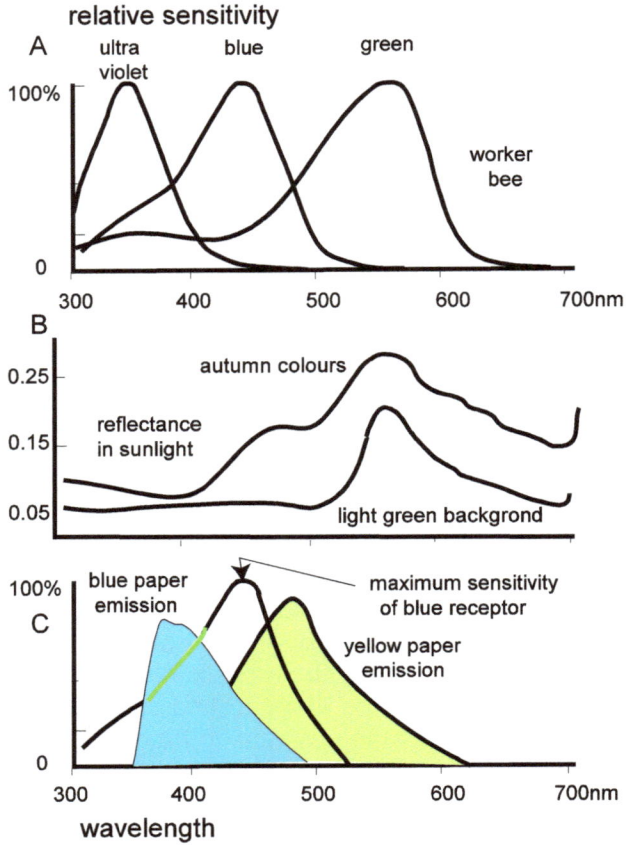

Figure 2. A. spectral sensitivity of the three types of photo receptors in each ommatidia of the worker bee retina, normalized to 100%. Actually the UV receptor has the greatest absolute sensitivity. B. Typical distribution of spectral emission of common scenes. C. The blue channel cannot distinguish deep blue and yellow and many others not shown, that emit the same total stimulus to the blue receptor, as shown by the equal areas of overlap within the spectral sensitivity curve of the receptor with a peak in the blue, near 450 nm. This is the definition of equiluminance to the blue detector channel.

Design Requirements Drove Evolution

To explain the vision of the flying bee in a holistic rather than analytical way, I begin where no scientist would. Vision in insects is essential for control of flight. Consider first the visual performance needed by a flying insect that regularly returns to a flower bed for nectar and pollen, and then returns home after each visit. To do that, bees do not need to distinguish, categorise, and learn objects of different kinds, and later recognize them as humans do with their large brain. They do not even recognise their own hive. If it is moved a little they fly to the former place where they expect the entrance to be. The worker honeybee, which is a sexless herbivore, needs to distinguish only those places along the route where a change of course is required. This implies learning and recognition of simple signposts that have nothing to do with flowers or any particular object. Anyway, flowers are quickly and more conveniently recognized on arrival by their specific odours.

Compound Eyes

Honeybees see directly forward but look mainly to the side, and use optic flow all around to measure range and distance travelled. The facets of the eye are very small, only about 26 microns diameter (1 micron = 1 thousandth of a millimetre). Each is a lens with its focus below on the end of a light guide containing rhodopsin, a visual pigment in a rod like our own. A narrow beam of light about 2° wide, is absorbed into the pigment, causing a lowering in the small potential, 50mV, across the receptor cell membrane. This potential change is the first signal in the nervous system. It spreads along the axon of the receptor cell to synapses on the first layer of neuron processing as a smooth voltage change without nerve impulses. This unusual transmission is essential because the signal is measured quantitatively and smoothly over a wide range. In all animals with a central nervous system, critical processing and initiation of behaviour is done with this kind of transmission. Axons are for crude rapid distance transmission by nerve impulses that are not so graded, involved in inhibition, or extremely interconnected.

Receptor Inputs

Only two of the three colour types of the retina (Figure 2A) are used to detect features in the visual field. Foliage reflects little UV, but the UV receptors are mainly involved in the escape response to the sky. The blue receptor is so called because the peak sensitivity is in the blue of the spectrum. As demonstrated by testing bees with the head clamped, the only effective steady illumination is the colour blue. The blue receptor is sensitive to the fraction of blue emission from areas of any colour, and also to edges with contrast to blue receptors. For example, yellow emits about 10% of the blue emission of white paper, and green about 50%. Different colours can be distinguished, as colour-blind humans do, by different amounts of blue relative to green foliage in the background. The green channel is sensitive only to contrast at edges, not colour of areas. To be detected, all contrasts at edges must move relative to the eyes. Bees scan to left and right in the horizontal plane as they fly, so vertical edges give the strongest stimulus.

Contrast and Modulation

Human vision sees the panorama of the outside world as a wonderful spatial projection upon an internal visual system, and keeps separate the edges and contrasts across edges. Honeybees scan from side to side as they fly, and their vision is quite different. In the blue channel, the responses to blue emission are measured and simply summed over large regions (or all) of each eye to give a running total of quantity of blue, and its height above the flight line. This continually changing pattern of blue input can be learned like a tune or a program as the bee scans, and helps recognition of place, but does not detect detail.

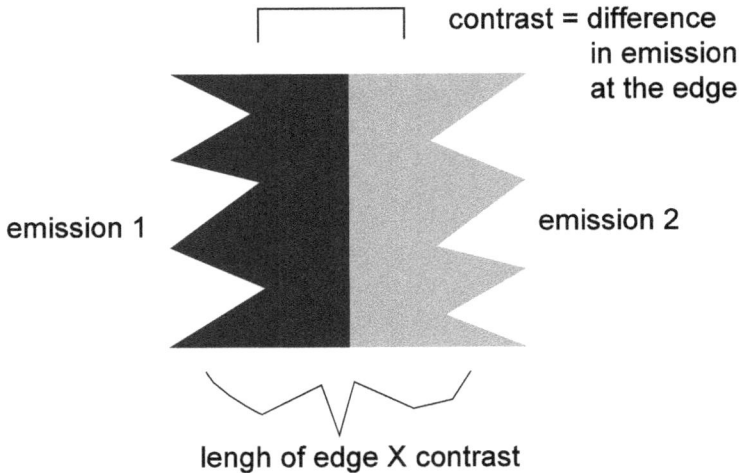

contrast = difference in emission at the edge

emission 1

emission 2

lengh of edge X contrast

Figure 3. The definition of contrast and modulation. Contrast is the step in effective emission at the two sides of an edge, measured as (e1 minus e2)/(e1 plus e2). Contrast is a stimulus outside the eye. In human vision, shape and size of an edge are distinct from its contrast, but the bee detects the change in response of the receptors inside the eye, called modulation, which is the amount of edge in the visual field of the receptor multiplied by the response to green contrast at every bit of the edge that is detected. Modulation therefore includes an aspect of the structure of the visual field, not just contrast.

In the green channel, contrast and the length of edge at each contrast outside the eye are detected together and not separated, so responses inside the eye include the amount of edge as well as its contrast (Figure 3). Length of each piece of edge multiplied by the contrast at each bit of the edge is called modulation, and this is the input that is measured, summed and learned over large regions (or all) of the eye as the bee scans. Moreover, lengths of vertical edges are emphasised by the scanning action. Because modulation is a product that cannot be turned back into the original spatially distinguishable edges, bee vision is unable to distinguish shapes.

Eye Movement is Essential

Not all features available to humans are detected by bees. To detect signposts and landmarks irrespective of range, flying insects require mechanisms that work irrespective of apparent size, otherwise it becomes very laborious to learn and remember the signal of each signpost or land mark as seen at a different range as the bee flies (Figure 4). This requirement for a repeatable recognition signal independent of range limits the variety of features that bees use. Humans do better with binocular vision.

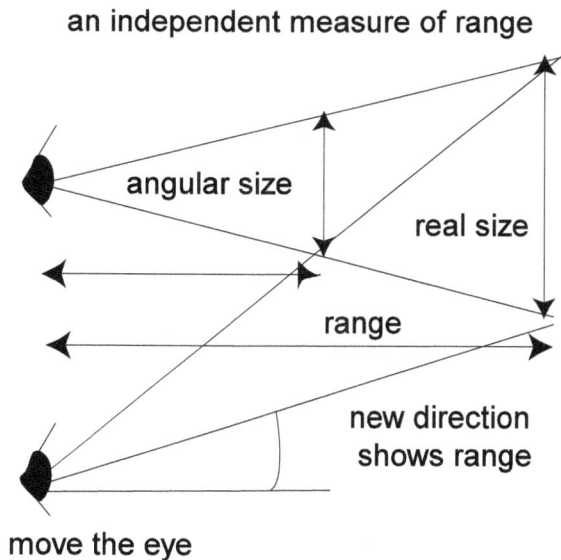

an independent measure of range

angular size

real size

range

new direction
shows range

move the eye

Figure 4. Eyes function in radial co-ordinates with the eye at the centre, so the image is an angular size, but the true size depends also on the range. Therefore, to obtain true sizes, additional information is obtained by detecting motion of green modulation relative to the moving eye, called optic flow.

Training Bees to Choose Between Targets

When they are trained to select one of the two targets in a simple binary choice (Figure 5), they first learn to avoid the unrewarded target (See Figure 11) because that is the one that tells them they are in error, whereas if they arrive first at the rewarded target, they leave with a reward and have no reason to learn anything.

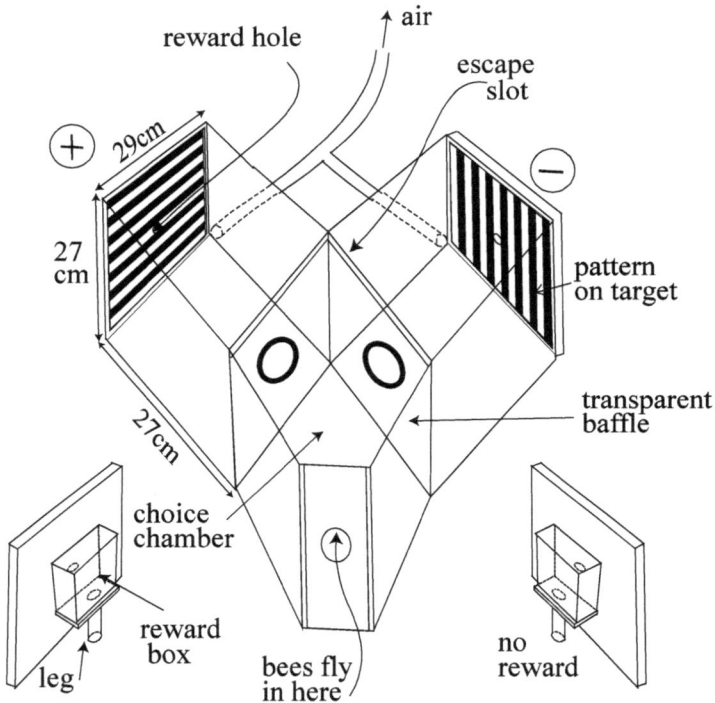

Figure 5. The Y-choice apparatus used to train and test bees. The top is of thin transparent polycarbonate, the sides of wood covered with clean paper. Transparent baffles force the bees to stop and look at the targets. A flow of air through the apparatus prevents a spread of odour clues. The reward box contains a feeder with dilute sugar solution. The movable targets, together with the reward, are interchanged every ten minutes during training. This apparatus allows complete control of what the bees may see, and excludes all except the marked bees. Bees take about 20 visits to learn that the unrewarded target offers them nothing.

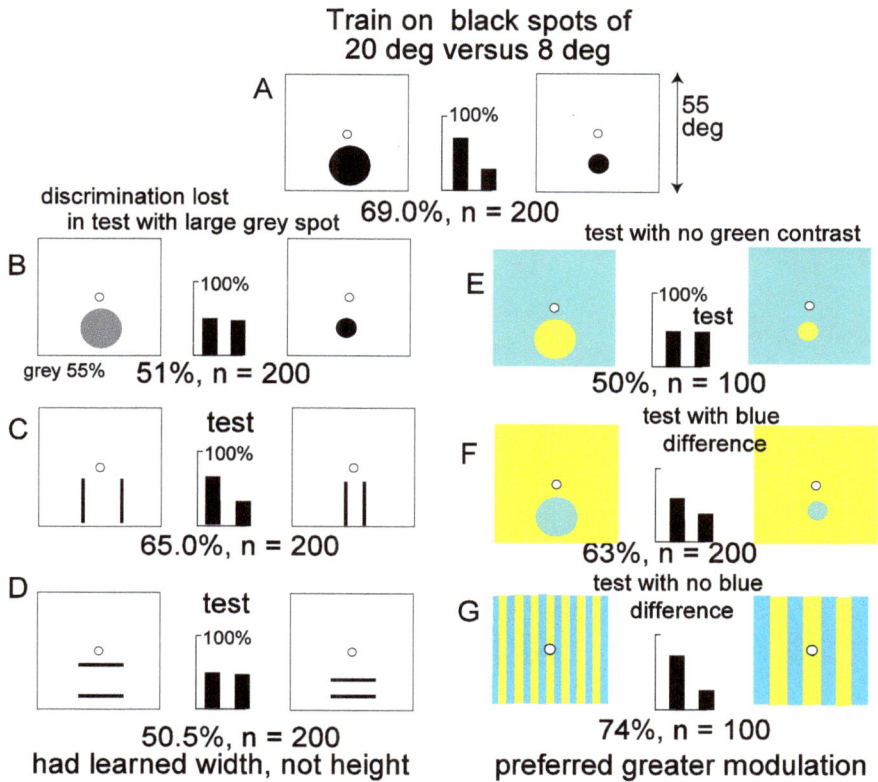

Figure 6. An illustration of the use of the Y-choice apparatus for training and testing bees, to solve Ronacher's anomaly. A. Bees were trained to distinguish a large black spot from a small black spot. B. When tested with a grey large spot, the trained bees failed to distinguish the spots. C, D. They had measured the width, not the height. E, F. When tested with a small blue difference, the response depended on the relative amounts of blue in the test targets. G. They preferred the greater green modulation, learned from the extra edge length of the large black spot.

Eight Features that Bees Distinguish and Learn

The following features in the environment that bees distinguish and use for route finding were all discovered in the 21st century. It is important to realise by the time you reach the end of the list, that they are all generalized relationships that are not colours or shapes, and are little influenced by changes in range. When making a choice, bees learn or recognise relative measures, not absolute values (Figures 7, 8).

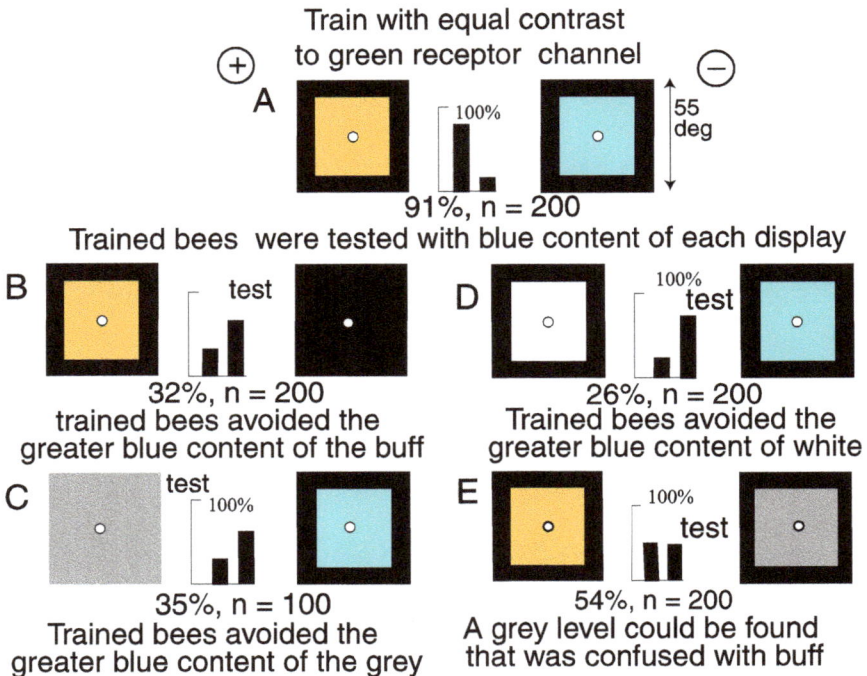

Figure 7. A. Training patterns. B, E. When tested with two patterns displaying different blue emission, the trained bees avoided the target with the most blue, showing that they used the relative difference, not the absolute values of the emission in the training.

1. A Quantity Surveyor

The total quantity of blue emission (Figure 7) or green modulation (Figure 8) is detected and summed over the whole eye, so the fixed field of view determines the internal response, and provides optimum certainty of recognition when detected and measured again. This measure is a characteristic of the place. Even when nothing else is detected, a total with no visible polarity may indicate a familiar place, like heading on a straight course towards a flower with bilateral or circular symmetry. This system has poor spatial resolution for recognition of detail because the blue emission and green modulation are separately summed over the whole eye.

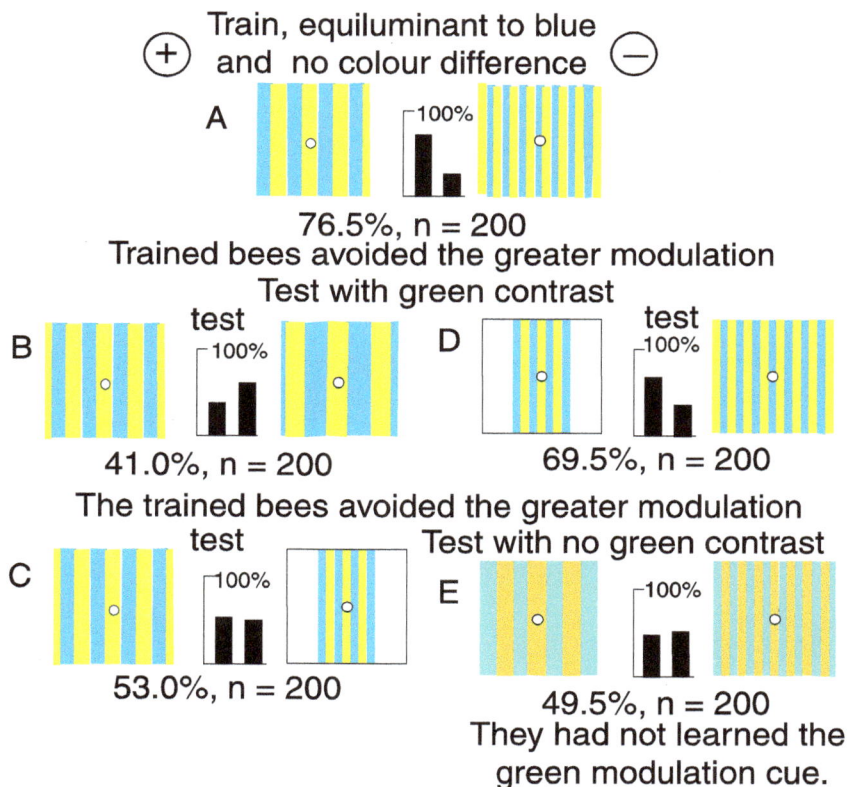

Train, equiluminant to blue
(+) and no colour difference (−)

A

76.5%, n = 200
Trained bees avoided the greater modulation
Test with green contrast

B
41.0%, n = 200

D
69.5%, n = 200

The trained bees avoided the greater modulation
Test with no green contrast

C
53.0%, n = 200

E
49.5%, n = 200
They had not learned the
green modulation cue.

Figure 8. A demonstration of learning a relative difference in green modulation. A. Bees were trained to avoid the greater green modulation with no difference in blue content. B − D. In tests, they avoided the greater modulation and ignored the pattern. E. With no green contrast, the trained bees ignored the patterns and failed.

2. Signposts Are Not Range-Sensitive

A signpost must be effective irrespective of range or illumination. The left/right arrangement of green modulation relative to blue colour, like a mirror image, is a preferred signpost for bees. The most common is the relation between a patch with some blue emission on the left or right of a vertical edge of green modulation that acts as a landmark (Figure 9). These directional inputs are signposts reminding the bee to turn left or right irrespective of range. The bees do not 'see' signposts, they scan and detect their polarity. Recognition may be aided by an additional measure of total modulation emission at that place. This system has a high spatial resolution, similar to the angular separation between facets. These signposts, of different sizes, occur all along the bees' route.

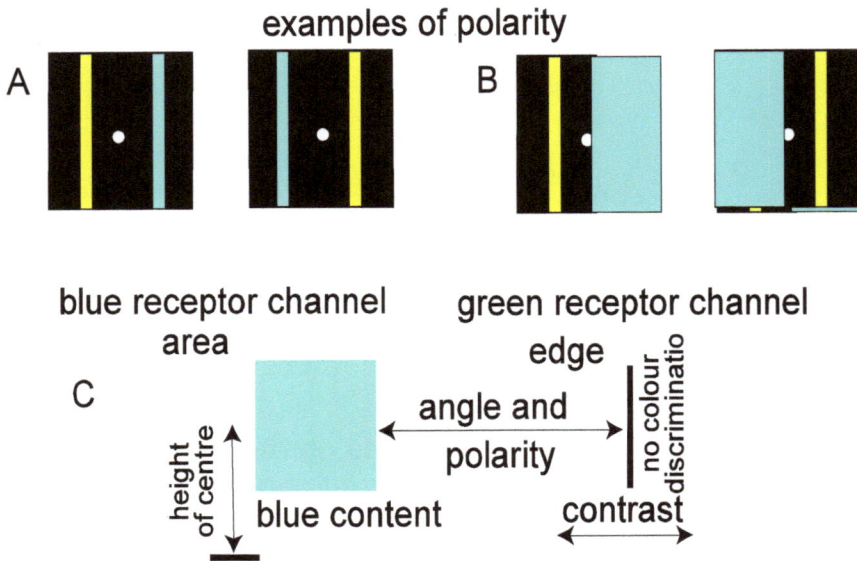

Figure 9. A preferred feature that bees discriminate is the left/right spatial relation between a vertical edge of green contrast acting as a landmark, and the average position of the blue content of emission of an area that might be blue, grey or any other colour. The signal is the direction of the polarity and relation to blue content, not shape, size or position.

3. Measurements of Intervals

Vision by scanning enables the bee to make measurements of time intervals between bursts of modulation. This ability, quite unknown in human vision, enables them to detect and measure and identify gradients of green modulation (Figure 10A). The left/right direction of the slope of modulation intensity is detected, and used as a signpost with polarity (Figure 10C). The bee also has some measure of changes in time elapsed between sudden bursts of modulation as it scans, as shown also by the ability of bees to distinguish a width between vertical edges, and to learn particular widths of vertical bars or tree trunks. This is part of the mechanism with which bees learn to distinguish widths of objects irrespective of distance, and range irrespective of size. Other simple temporal sequences of modulation emission are distinguished, learned, and later recognised, as happens in insects that fly a familiar route, detect particular sequences of clicks in their song, or gradients of odours as they fly across wind. Humans also readily learn and recognize temporal patterns in song or dance, but we have no idea how it is achieved.

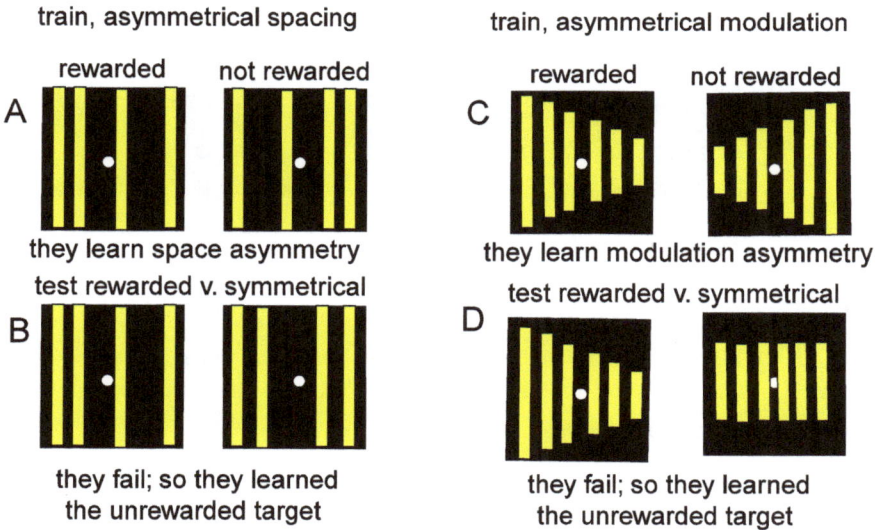

Figure 10. A, B. Bees can be trained to discriminate cues from the delay between successive modulations as the bee scans. C, D. Bees also detect gradients of modulation and their polarity as they scan. The use of yellow on black helps to avoid a blue difference.

4. Tangents and Spokes

In the 1930's several authors reported that honeybees distinguish radial patterns of spokes from concentric circles (Figure 11A), and both of these from other patterns that lack these features. For a reward, honeybees certainly learn to fly through a small hole, although normally they will not, and it is difficult to persuade them to fly through wire netting with a mesh of less than about 2 cm, and there are many reports that they avoid spiders' webs. In general, they avoid circles and prefer a set of six spokes separated by 60° (Figure 11C), but dislike right angles and fail to distinguish a + from the same rotated by 45°. Recognition of these features seems to be related to the hexagonal pattern of the facets on the eye.

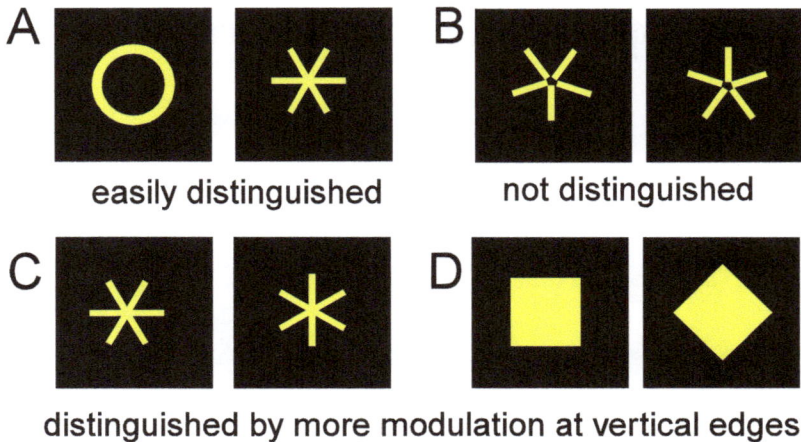

A easily distinguished   **B** not distinguished

C **D** distinguished by more modulation at vertical edges

Figure 11. A. Bees discriminate between circular and radial edges, especially if symmetrical, B. They find it difficult to discriminate between circular patterns or radial patterns of similar size. C. They easily distinguish radial patterns of six equal spokes. D. Rotation of a square is distinguished by a difference in total modulation, mainly from vertical sides.

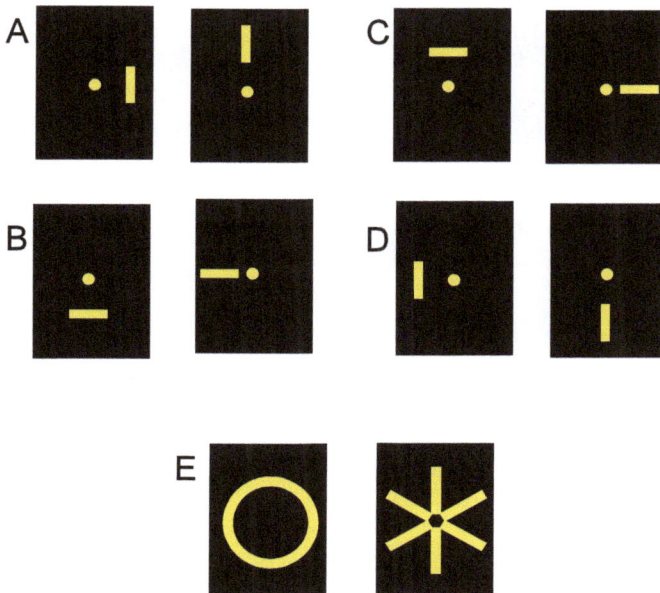

Figure 12. A. D. Bees were trained to distinguish between two yellow spots, one adjacent to a yellow tangential bar and the other near a radial bar. The targets were rotated through 90° round the spot every 10 minutes, to randomise their positions but retain the relation of the bars to the spot. E. Tests showed that the bees had learned to distinguish tangential edges from radial ones.

Bees are spontaneously attracted to contrasting small spots, and detect them on a mixed background. Nearby edges or lines in the background are treated as if they are radial or tangential relative to the spot (Figure 12). This suggests that there are dedicated feature detectors for radial symmetry, as in flowers. As bees evolved detectors for asymmetry, flowers evolved symmetrical flowers from the spiral apical meristem, with the happy result that flower forms do not signal to arriving bees to divert to the shop next door. Flower forms have different modulation patterns.

5. Edge Orientation in a Vertical Plane

In the 1930's it was discovered that the orientation of a line or edge in the vertical plane was discriminated and learned. With a group of edges displayed in the vertical plane, bees detect, learn and recognise the average orientation (relative to the vertical). With their green modulation channel, they also measure the total green modulation within large solid angles that can be smaller than the whole field of each eye (Figure 13B). Edge orientation is their least preferred feature.

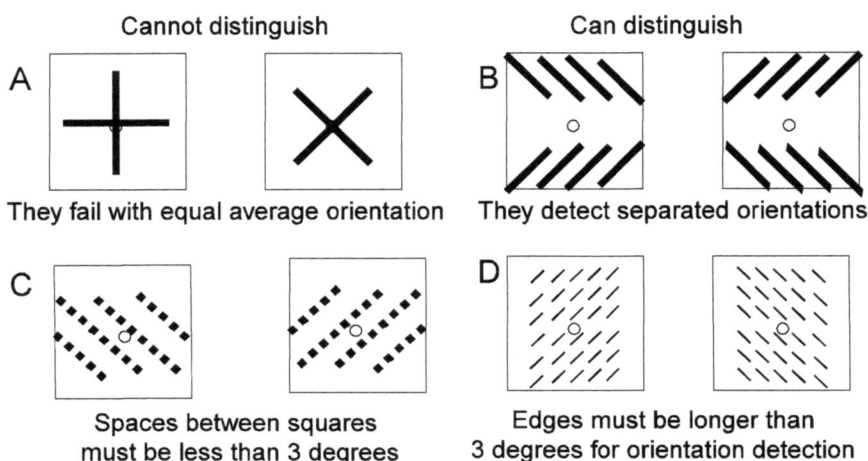

Figure 13. A. Orientation of edges in simple targets is summed in such a way that edges of equal length at right angles cancel the orientation cue but retain the modulation cue. B. Well separated strong orientation cues are detected separately. C. To be detected as an orientation cue, small squares in lines must be separated by less than 3 degrees, and D, short lines must be at least 3 degrees long. When training with black patterns on a white background, a possible difference in blue content of the background is avoided by the use of exact mirror images.

Parallel gratings of different orientation in a vertical plane are popular targets that are easily discriminated, although they are the least preferred cues when others are available. Later it was shown that equal lengths of edges or lines at angles to each other mutually cancelled the orientation. Equal lengths of similar edges at right angles cancelled (Figure 13A). The effect is reduced when the different orientations are spatially separated by more than 20°, which is less than the field of the whole eye (Figure 13B). Average orientations of groups of edges separated by angles of 20°or more are distinguished. For example, one group of parallel edges like grass stems can be distinguished from another group at a different orientation if they are separated vertically by more than 20°. These averages and totals of the same features are identification cues of particular places. Tests of the minimum size of the orientation detector reveal a length of 3°, i.e., 3 facets (Figure 13C, D). Detectors along an edge are not strung together as they are in humans. This short baseline (three facets) explains why the lower limit of bee discrimination between two oriented edges is poor, about 15°.

Only the green receptor channel detects edge orientation (Figure 14). Therefore the blue receptor channel cannot detect pattern.

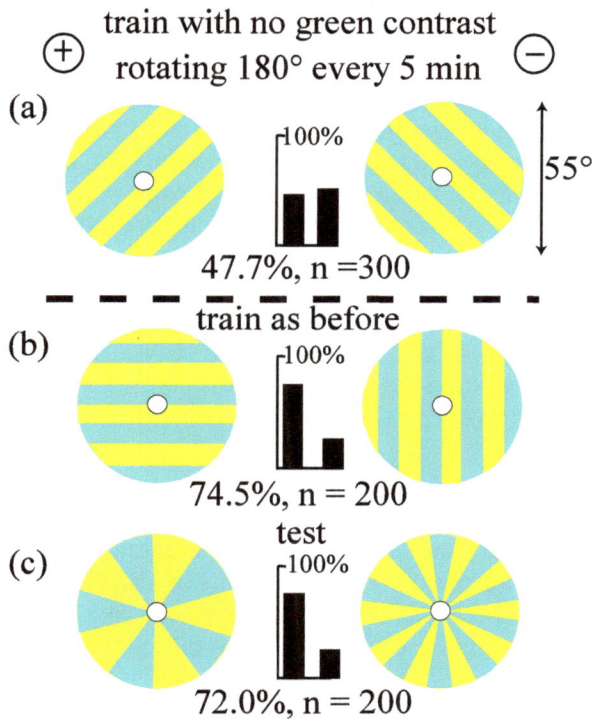

Figure 14. With no green contrast, the two oblique gratings were not distinguished. B, C. When trained with horizontal versus vertical bars, bees distinguished the relative difference in modulation between quite different patterns.

6. Local Motion Detectors

Many insects have a separate system of optic lobe neurons that respond to a small target with green contrast that is detected against a moving background. Bees are no exception; clearly, in flight they detect other bees and avoid small twigs in their flight path. Robber flies and dragonflies catch their prey in flight; similarly, mayflies and drone bees catch their mates in mid-air, and mantids grab accurately at moving prey. Research on these and similar behaviour suggests neurons that detect the direction of a small object of the right size that moves exactly like the prey or mate in the expected situation in any direction relative to the eye. Recordings of these neuron responses are well known for dragonflies and mantis.

7. Station-Holding and Motion Measurement

A separate system of green sensitive motion detectors, with high spatial resolution, not related to recognition, detects the motion of green modulation from one facet to the next across the eye, and sends a strong response into wide-field neurons that detect motion and its direction, to hold the position in space, and control flight direction, speed, and flight posture. This colour-blind system also detects the motion of the whole optic flow, and is able to provide a measure related to the distance travelled, as long as the route remains reasonably homogeneous. This is how bees measure and learn the distance that is conveyed to other bees in the dance. Bees can hover relative to a nearby object and also can fly across the wind direction, by reference to the ground, as they search for odours.

8. Compass in Ultraviolet of the Sky

Bees are sensitive to and sum together the plane of polarization in the ultraviolet emission across large regions of the sky. Bees have a sense of time of day, and know the position of the sun in the sky. They learn this data, like a table for navigation, and remember the direction of home at all times.

How Widespread is this Type of Vision?

All endopterygote insects (Insects that have a maggot or grub) have a system of visual receptors and optic lobe neurons similar to that listed above, which suggests they have similar visual systems. Many insects return to familiar feeding or resting places. Worker honeybees have only three colour types. In some flies and butterflies there is an additional red or yellow-sensitive colour channel that seems to be associated with specific identification of that colour in mating behaviour or recognition of food or prey. This is not true colour vision. Von Frisch claimed to demonstrate discrimination of hue of colours by the worker honeybee, but careful study of his publication shows that his bees could not distinguish various green papers from mid-grey, and they summed together each array of 15 grey levels that he presented together.

The general principle of bee visual recognition is that bees take a quantitative measure with all the different feature detectors as they scan the scene, and they are able to learn and recognise again the unique combination of several totals at each place. The more detectors are active at each scan and the more scans, the better the recognition. Identification of these detectors by testing trained bees is physics or physiology not behaviour. The insect visual system does not even detect the panorama of things that humans distinguish and categorize; it is designed to detect polarity of signposts and several general properties of the scene that will function irrespective of range.

For all Detectors, Input Resolution is Essential

We might ask why there are so many facets in the eye, each 25 microns diameter (1 micron = one thousandth of a millimetre). The great number is possible because the wavelength of light is so small, half a micron. The smaller the facets, the more there are to separate more edges by spatial resolution. But more facets implies a smaller aperture of the lens, so sensitivity and resolution of detail are lost. To function well in daylight, each ommatidium must have an F-number that lies between 3 and 10, like a camera or the human eye. The design of the eye is optimum for a combination of sensitivity, point resolution, and resolution of separation of edges, but the small aperture limits sensitivity, so that bees prefer to work in sunlight and must finish work early.

Genetically Determined Machinery Must be Taught

Foraging bees have no independent prior knowledge of the inputs their complex environment presents, but their feature detectors have evolved to fit the task, and are genetically determined. They detect sums of modulation, left/ right polarity, and polarized gradients of intensity or modulation. These are all features that natural selection had isolated from the emission of the panorama because they were to some extent independent of range. Insects with ears are able to learn and recognise particular sequences of the specific songs of their mates as patterns in time. In addition to familiar odours learned while nursing larvae, foragers detect the features listed above, and the sequences in which they are detected as they fly. Spatial frequency of edges (green modulation) can take many forms, from noise with many spatial frequencies, as in bushes or forest, to an isolated modulation, as when passing a tree trunk. As they fly, they they detect the features and signposts that they need, and they can learn a long symphony of cues and chords.

Do Bees Think?

We know a lot about the features that bees detect visually in flight because the physical properties of the photo receptors (Figure 1A) have been directly measured with micro electrodes, and emission from targets and panorama (Figure 1B, C) can be measured directly with a photometer. We can control every aspect of the patterns to be learned, and measure the responses of our trained bees to as many test patterns as we have patience to present. Bees do not 'see' the panorama, visual signals pass from the optic lobes behind the eye into neurons to the motor system, labelled with signals like stop, turn, or fly faster, We know nothing about deeper neural mechanisms involved in how bees make choices, acquire a memory, or the process of recognition. Mechanisms and control of more complex behaviour, like care of brood, making honey, or comb building is an even greater mystery, all done in the darkness of the hive where odour is the dominant sense.

The reason that we are ignorant of how central nervous systems work is that Anatomy tells us only the geometry of the nerves, not their Connectivity. Recording from them, even individually, tells us only when these structures are active. The meaning of a message is conveyed by the identity of each line, which is chemically labelled via the developmental mechanism based on the DNA of each separate neuron. It is not obvious that many neurons are inhibitory, not excitatory. Apart from the peripheral sensory nerves, only rarely do we know what stimulus each neuron conveys. The context of each with other neurons is even more elusive, but methods are being developed to reveal which neuron groups have been active together. Human language is arbitrary and our dialect fools us into attributing reality to ideas. If in doubt, I advise much deep thought.

If you have few years for the proper use of leisure, you might be interested in extending these findings to other animals. What does your pet dog, cat, budgie, tortoise or goldfish really detect with its eyes? At present we know a little about their capabilities and performance in visual tasks, but have no information at all about the feature detectors or other mechanisms that make the marvel of vision possible.

REFERENCE
References for all details in this account will be found in my recent book:-
Horridge, Adrian. The Discovery of a Visual System.
CABI Books. Boston, USA and Wallingford UK. 2019

www.ingramcontent.com/pod-product-compliance
Lightning Source LLC
Chambersburg PA
CBHW041430270326
41934CB00020B/3494